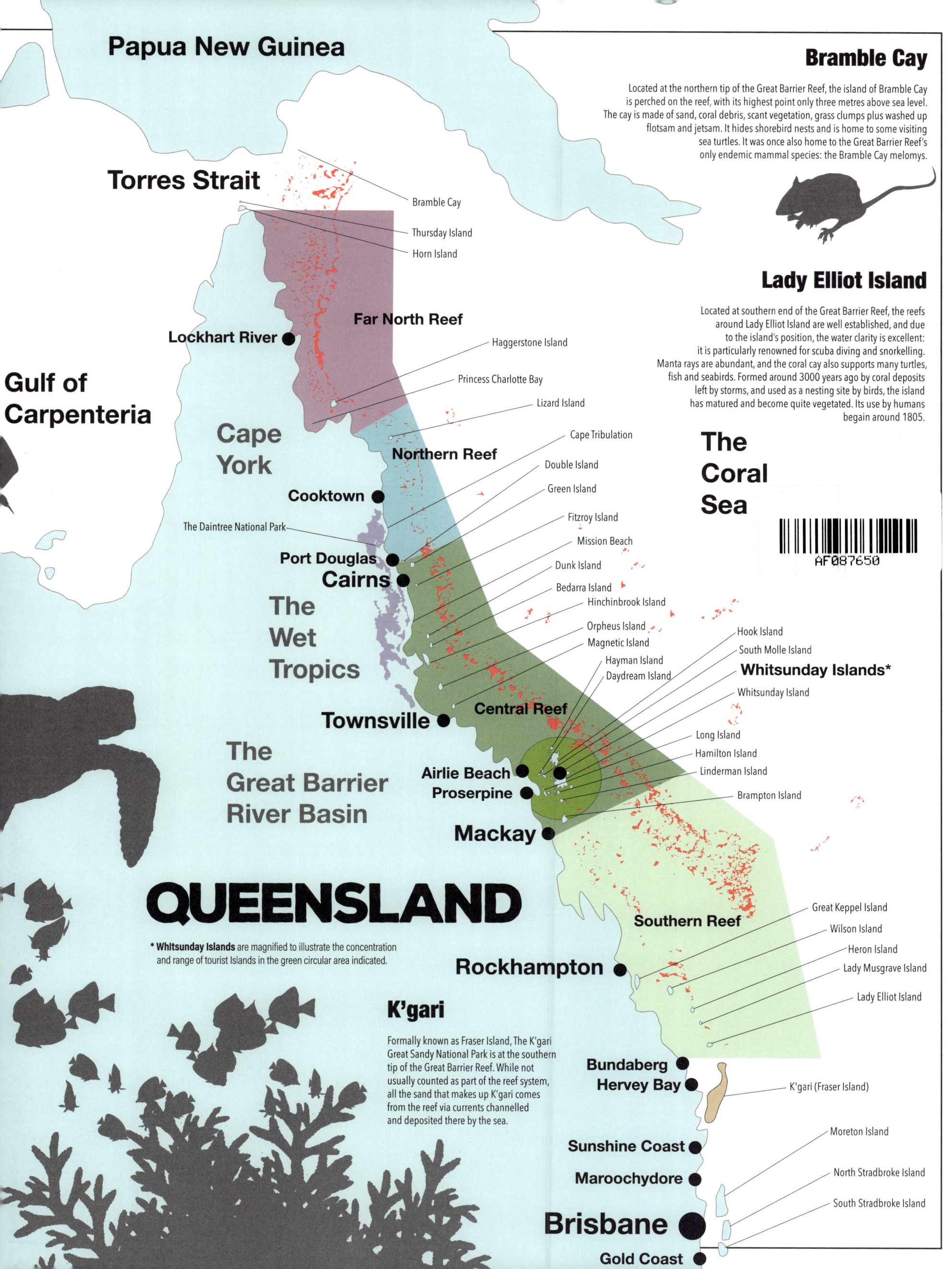

Wildlife of Australia's GREAT BARRIER REEF

Myke Mollard

QUEENSLAND'S MARINE EMBLEM

The clown fish immortalised by the popularity of *Finding Nemo* still enchants us with its character and colourful markings. Darting out of sea anemones on the reef, these highly territorial fish, no bigger than the palm of your hand, display a skittish behaviour. One moment bold as brass and the next moment retreating to the safety of the anemone. It's comical to watch and hence their name, clown fish.

This book is dedicated to all those people who cherish the ocean and the majestic beauty of what lies beneath the waves. Especially the next generation of divers and adventure seekers like my kids: Xavier, Benjamin and Annabelle.

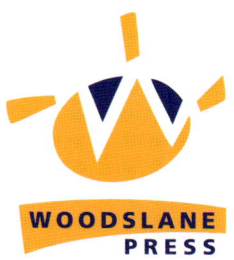

Woodslane Press

A HEALTHY CORAL REEF
No matter how you see the Great Barrier Reef - from the air, while sailing through coral passages, looking down while snorkeling or scuba diving deep - you are likely to have your breath taken away by the seemingly endless beauty of the scenery.

LET'S DIVE ONTO THE REEF

A barrier reef is a long coral formation that lies parallel to the shoreline. Mostly underwater, its structure and formation is quite different to a fringe reef or coral atoll. Located off the northeastern coast of Australia, the Great Barrier Reef is the mightiest of its kind in the world, the earth's largest living structure and the most widespread connected coral reef of any ocean. Its northern reefs are part of the Coral Triangle, a system of reefs stretching from the Philippines, Malaysia, Indonesia and Papua New Guinea out across the Coral Sea to the Solomon Islands and arguably as far as Fiji.

The Great Barrier Reef can be seen from space - it's that enormous! It stretches from Bramble Cay, in the Torres Strait, to the unnamed passage between Lady Elliot Island and K'gari (Fraser Island) in the south. That's over 2,300 kilometres long, encompassing over 344,400 square kilometres. This World Heritage Area comprises over 2,900 individual reefs and more than 900 islands dotted throughout. The Great Barrier Reef couldn't be a more apt name. It is separated from the Queensland coast by a channel (in parts this passage is 150 kilometres wide, in some places over 70 metres deep). The reef shelters the Queensland coast and creates a marine home for billions of creatures.

So where did this extraordinary natural wonder come from?

Around 25 million years ago, what is now Queensland was still south of the tropics, but as the continent drifted north it moved into more tropical, sunlit waters. This enabled and encouraged coral growth, which then expanded and declined with the rise and fall of changing sea levels.

Between 24 million and 10 million years ago, the Great Dividing Range eroded down quite energetically, the detritus smothering many early reefs within this marine area. Mud, sediment from river deltas and land that crumbled into the sea or was shaped by waves helped shape and shallow the already extensive coastal plain.

The global ice ages of the last 3 million years have been a challenge to these reefs, lowering the sea level, draining the coastal plain and so killing any corals and creatures. During the interglacial periods the plain flooded once more and reefs were able to re-establish - a particularly extensive reef persisted for many millennia 400,000 years ago.

The last ice age waned around 20,000 years ago, the coastal plain inundating again. With rising sea levels, corals invaded once more, growing higher on the newly submerged hills and protected from rough seas by a chain of newly formed continental islands. Most of these continental islands have since eroded, leveled by the sea. By 6000-3000 years ago the reef was very much how we see it today.

The Great Barrier Reef supports an extraordinary diversity of life. As well as the coral itself and other marine life, the reef's tenants above the waves are not inconsiderable, and include invertebrates, birds, mammals and reptiles, all making a living in and on the many islands scattered throughout the reef. While some of the reef species are globally widespread, a significant number are endemic.

It's important to recognise that the reef itself can be thought of as a living structure. Corals are animals that stay rooted to one spot and as generation succeeds generation their skeletons gradually build the reef's architecture. They are at the centre of a web of diversity that encompasses a huge range of animal species.

There are myriad relationships between all the corals, fish, turtles, birds and other creatures in this ecosystem, all helping to reinforce the stability of the whole. Learning about these connections, behaviours, migrations and roles makes researchers even more curious. Many people see the reef simply as a resource for things like money, tourism and medicine, but it is so much more than that. We now know the reef performs unseen environmental functions that are of incomparably greater worth, such as a nursery for countless oceanic fish and other animals, and as a simply colossal carbon sink, helping soak up some of our excess carbon emissions.
We are beginning to understand that helping to keep this alive is as important for us as its denizens.

SEA FANS AND WHIPS
Sea fans, sea pens, sea whips, sea plumes, sea rods and feather stars all take on these beautiful fern-like structures.

SOFT CORALS OF THE REEF
Below are some of the soft corals that you will find clustered and dotted throughout the reef illustrations in this book.

SOFT TREE CORALS
Coronation coral tree, soft tree corals, flower coral trees, and tubual coral trees, branch out like great underwater bonsai trees.

POLYPED SOFT CORALS
Sun coral polyps colonies and many colourful flower-pot coral.

TRUNKED FINGER CORAL
Soft finger leather coral tree looks like an anemone but are quite different.

TUBUAL SOFT CORAL
Organ pipe coral colonies grow steadily in clustered tubes that resemble piping.

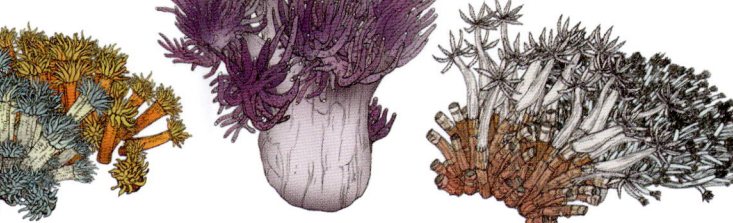

OUR UNDERSEA CORAL WONDERLAND

The number of coral species worldwide is not even loosely known, but there are probably many thousands. Corals are divided into two camps: the hard, reef-building corals, of which nearly 600 species are known on the the Great Barrier Reef; and the soft corals. The former build calcium skeletons which - in their millions - slowly accumulate into heaps and hills. The latter are the ones usually found in aquariums and many species are often confused with their close relations, the anemones. The corals of the Great Barrier Reef are incredibly diverse and environmentally flexible, reaching from the low water mark along the mainland to 250 kilometres offshore, and from the shallow inshore areas to the mid-shelf and outer reefs, even down to the base of the continental shelf. Over several millennia, thousands of generations of corals have built 2,900 individual reefs and islands, ranging from small sandy cays to larger vegetated islands. Collectively these marine landscapes constitute a spectacular 'coral rainforest'.

Hard corals are the architects of the reefs, vast communities of tiny creatures ('polyps') that create extraordinary stone-like structures as external skeletons or homes. Most species are, to some extent, solar-powered! Although they capture and eat a variety of zooplankton, their outer layers play host to algae ('zooxanthellae') which photosynthesize sunlight to create oxygen, carbohydrates and other minerals, some of which, in exchange for a safe home and carbon dioxide, are passed on to the host. Both partners pass essential nutrients to the other in an ingenious example of symbiosis. It is the pigments in the zooxanthellae that give corals their beautiful orange, red, purple, yellow and other colours.

While hard corals have a cup or tube-shaped outer skeleton, soft corals have internal skeletons inside their bodies, though these are not hard calcium structures but have a more pliable, woody texture. All corals grow close together, even on top of each other.

A great many marine species depend on coral for food and shelter. For many creatures, such as fish, octopuses and crabs, coral provides dens and places to hide from large predators. More sedentary creatures such as sponges and anemones get a foothold in between the hard and soft corals. All these in turn provide food for other creatures such as larger fish species and turtles - and so the ecosystem expands and becomes more complex.

Two animal families in particular sometimes get mistaken for corals: sponges and anemones. Sponges have bodies with intricate skeletons, but made of silica not limestone: there are even rare places in the world where they form reefs. Some anemones are assumed to be soft corals, but they they often kill coral and colonise with their stinging tentacles.

THE CORAL CLUSTER
Like a densely populated undersea rainforest, the coral reef is a huge, tight-knit community of organisms. Most coral species form colonies, with the individual polyps sharing a limestone skeleton. Most are filter feeders who catch tiny floating animals called zooplankton.

ARBORESCENT **CAEPITOSE** **CORYMBOSE** **DIGITATE** **ANTLER** **TABULAR**

Branching extending corals — Radial extending corals

UNDERSTANDING THE HARD CORAL STRUCTURES
Coral structures help us identify the different species of corals, both hard and soft, as well as sponges and sea anemones, which together form a clustered community on the reef. The hard corals are the real reef builders and the soft corals, sponges and anemones compete for real estate in the spaces between.

   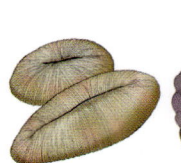 

COLUMNAR/FINGER **MASSIVE** **FOLIOSE/LAMINAR** **SOLITARY/FREE-LIVING** **ENCRUSTING/LOBE**

Vertically extending corals — Horizontally extending corals

THE GREATEST LIVING CORAL GARDEN
When you slip beneath the waves and float down onto a healthy coral landscape it's like nothing else on earth. The shallow coral is a kaleidoscope of colour, teaming with fish darting in and out of the coral fingers. Every inch of limestone is covered like a dragon's treasure horde, every nook and cranny filled with creatures struggling to keep their foothold on the reef. And it's noisy! The immediate impression of quiet is soon broken by the click and chatter of the fish as they feed.

THE FANTASTICAL COLOURS OF REEF FISH

When you dive onto the reef you will come across a deep rainbow of colours. It's not just the coral, it's also the marine life, especially fish. Ten metres down, everything to our eyes turns a deep blue as we lose the red end of the spectrum. But switching on a dive light reveals a wonderland of colour like nothing else on earth.

A coral polyp gets its vibrant colour from the algae in its body, while fish basically get their colours from their diets. This might explain why reef fish - those eating coral and sponges - are generally more colourful than predatory reef fish, estuary, pelagic and ocean-travelling fish. Fish colouration is thought to have four purposes or functions: thermo regulation (darker colours absorb heat better); communicating with shoal mates; communicating to different species (that they are bad eat); and camouflage on the reef itself. So whether warding off predators with colourful displays, hiding from predators in the nooks and crannies of the reef, or attracting a mate to raise little fish with, colour is mostly pure evolutionary advertising: a billboard coded into the scales of every fish to help tip the balance of survival in each species' favour.

Many fish make their colour pigments by extracting various chemicals from the often toxic corals, anemones and sponges that they eat on the reef: vibrant blues, pinks and greens, yellows and reds; bioluminescence in some cases. The latter is often when mating, prompted by hormonal changes. Fish sometimes change sex and this can also prompt a change in colour in some species. These colour changes can be slow or fast. Slow changes are often observed in species that alter their markings from a juvenile to an adult. Faster changing colours can occur when a fish is stressed (for example many fish change colour when caught) or when changing camouflage to suit a new environment: such as changing from a mottled pattern to mimic coral to silvery reflective colours to imitate a sandy sea bottom.

THE FISH CLUSTER
Across these pages, especially **opposite** and on the **following spread**, are pictured the most common, the most sort after, the tastiest, the most prized, the most fascinating and even the most dangerous fish of the Great Barrier Reef. Believe it or not, only ten percent of the reef's known fish species are represented here!

THE SWEETEST LITTLE HUMBUG DAMSELS
Like 'nemo' fish, their sometimes next door neighbours, the damsel fish are some of the most striking on the reef. One of the most playful is the humbug damsel fish which pops out of the coral it inhabits to feed. At any sign of danger these tiny fish dive back into the safety and security of the sheltering fingers of the staghorns. These 'damsels in distress' are quite mesmerising and never fail to delight.

CORAL CITADELS OF THE DEEP BLUE SEA

The Great Barrier Reef is - to us humans - a kingdom of majestic beauty and wonder. It is a citadel of limestone and silica, regenerating during most interglacial periods, often with a very different cast of characters. A treasure trove of marine biodiversity, over 9,000 recorded species have been spotted there to date, including corals, fish, crustaceans and molluscs. Some inhabitants look to our eyes much as their ancestors did 500 million years ago - long-term reef dwellers - whilst others have evolved only relatively recently to find a niche on the largest living structure on earth. It's easy to cast these denizens into roles: the kings of the reef, the colourful court jesters, the swift moving hunters and the brooding leviathans that haunt the dungeons.

Along with coral trout, chinaman fish, red emperors and kingfish, the groupers are also prized game fish of the reef. Of the approximately 200 species of groupers worldwide the Great Barrier Reef is home to over 70. Groupers are predatory fish and the largest is the Queensland (or giant) grouper, growing up to 2.7 metres in length and weighing in at over 400 kilograms. On the reef size matters and territory is everything. These often docile but imposing fish have a drab, mottled colouring, consisting of mainly speckled greys and browns. Their usually placid nature makes them a treat to encounter while exploring the reef, but they also have a reputation for occasional nasty attacks on divers.

Red emperors, kingfish and other pelagic species patrol the open water, deep channels and passageways between the coastline and the barrier reef. In amongst the coral are hundreds of reef specialists: the brightly coloured parrot fish and mottled, often iridescent wrasse, leatherjackets, triggerfish and sturgeon. One of the largest and most iconic of them all is the Maori wrasse or humphead wrasse, reaching a size of 2 metres and weighing up to 180 kg.

Parrotfish and wrasse are some of the most important fish on the reef. They graze the algae and their poop supplies nutrients to many corals. Parrotfish are wildly colourful and go through many wardrobe changes throughout their life until sexual maturity. There are many other amazing things about these fish, but one curious adaptation is that they create a mucous bag to sleep in every night to protect themselves from parasites.

KINGFISH, COURT JESTERS AND DAINTY DAMSELS
This page: yellow-tailed kingfish patrol the open water, cruising by a coral drop-off with yellow pilot fish or baby golden trevally as an entourage, shadowing their every move. Below them a sunset parrot fish looks for a place to rest, while a bumphead parrot fish checks for predators. Damsels dance around the coral shelf as a green parrot fish grazes the coral bed.

CORAL LEVIATHANS
Opposite: Bull sharks might be the most feared beasts patrolling the islands, estuaries and coastline of mainland Queensland, but other behemoths lurk in the shadows of the reef. Giant groupers, Maori wrasse, potato cod and even coral trout can get to pretty mammoth sizes on the reef. While usually gentle in nature, these fish can have a short temper and you should always remember you are a guest in their territories.

SEA BIRDS AND OCEAN FISH OF THE GREAT BARRIER REEF
Sea birds such as Brown Noddies and Ospreys may hunt the surface, but under the waves are the real sea-going hunters. There is a channel, up to 150 kilometres wide, between the barrier reef and the mainland. This ocean corridor harbours many predator fish species, including marlin, sail fish, giant trevally, golden trevally, kingfish, barracuda, tuna, snapper and dolphinfish, and even rare creatures such as the ocean sunfish.

SHARKS: THE FEARSOME AND THE TAME

A great number of shark species call the Great Barrier Reef home. The most common sharks found there are the whitetip and blacktip reef sharks; also common are grey reef sharks (also known as bronze whalers), leopard sharks, epaulette sharks and the shaggy-mouthed wobbegong sharks. While snorkeling or scuba diving on the Great Barrier Reef you will likely have a great experience encountering these reef-dwelling sharks. There are a few other shark species that inhabit the warmer inner waters on and adjacent to the reef, and some of these can potentially be more of a threat to humans, but are rarely encountered. One interesting shark that periodically frequents the reef is the massive 15-metre-long plankton-hoovering whale shark; they are known to move through the passages and channels off the coast and outer reef.

There are 370 species of sharks known globally across the oceans and Australia is home to over half of these. It's a shame that this fact keeps many people with a fear of sharks confined to a boat or firmly on dry land when visiting the Great Barrier Reef. The warm waters of the reef actually support many smaller varieties of perfectly harmless reef sharks.

The Great Barrier Reef's location, 40-100 kilometres off the coast, actually makes it the perfect place to experience sharks safely and without fear. Nearly all of the reef sharks simply mind their own business. From wobbegongs, grey nurse sharks and lemon sharks to the more elusive leopard and reef-walking epaulette sharks, the reef is a treasure trove of species from this very varied family of creatures.

However, even if the Great Barrier Reef itself is relatively safe, it's good to have a healthy respect for these wild creatures, especially when snorkeling and scuba-diving the coastal waters off Airlie beach, Shute Harbour and the Whitsundays.

Four out of the nine known species of the distinctive hammerhead sharks patrol the reef, and one does grow to a threatening size. Tiger sharks usually hunt turtles and large fish; they are fairly uncommon and rarely seen, usually avoiding us humans, but locals will talk about them launching occasional ambush attacks thanks to the murky coastal waters. The same goes for the infamous bull sharks. These can also be a problem in coastal estuaries and harbours where visibility is very low. There are virtually no great white sharks in these waters, and those that are are rarely seen.

Grey reef sharks (bronze whalers), related to bull sharks, are fast-moving sharks, slightly smaller than the tiger and oceanic whitetips. They are inquisitive and do have the potential to cause problems, though attacks are almost unheard of. Whitetips are not generally found on the inner reef, they tend to hang out in the deeper waters of the Coral Sea. Roaming foragers, they cruise the outer reef system and hunt in open water.

I love sharks! Their grace and presence is always felt when they cruise by or you discover them resting on the sand under a shelf of coral or limestone. It's many a scuba diver's dive highlight to have an encounter with a member of this ancient family, which has persisted with a largely unchanged body plan for over 300 million years!

A HAVEN FOR LEMON SHARKS

An unusual deposit of almost pure white silica sand makes up Whitehaven Beach on Whitsunday Island, only a short boat ride from Airlie Beach. Whitehaven Beach and Hill Inlet form a protected nursery for many hatching lemon sharks. If you keep your eyes sharp you may also see young blacktip reef sharks, stingrays and even green sea turtles enjoying the crystal clear waters of the Whitsundays.

CARPET SHARKS ARE CAMOUFLAGE EXPERTS
In the shallow reef and inter-tidal zones of fringing reefs, often the crechés for small sharks, you may find the amazing, reef-walking epaulette shark. Hemiscyllium ocellatum is a species of long-tailed carpet shark found in shallow, tropical waters off the coasts of Australia and Papua New Guinea. It is nocturnal and has two large, distinctive black spots on either side, above its pectoral fins. In slightly deeper waters, hiding under ledges or on the sand, is another carpet shark and one of the largest - the spotted wobbegong. This shark has complex colouring, enabling it to camouflage itself on the reef, and tassels around its mouth and jaws (hence the name carpet shark). There are 12 species of carpet sharks and we have over half on the Great Barrier Reef. They aren't dangerous to divers or snorkelers but if rambling through a rock pool you step on one it will rightly defend itself.

SHARKS OF THE GREAT BARRIER REEF
Many sharks share the reef, the largest being the majestic whale shark which comes seasonally during April-July. Blacktip reef sharks, whitetip reef sharks, grey reef sharks and lemon sharks are all common hunters on the reef itself, while tiger sharks are occasional visitors. Hammerheads of all sizes and bull sharks hunt in the adjacent coastal estuaries, mangrove environments and deep channels.

FROM BOTTOM DEWELLING STINGRAYS TO MIGHTY MANTAS

There are over 200 species of stingrays worldwide. They vary in size and colour, the most common characteristic being the flat disk-like fins which act like wings. Their tails are usually long, slender and whip-like, but can be short, shark-like or with lumps or thorny lobes. In calm coastal shallows, in muddy or sandy estuaries, or in shoals across the reef, most species rummage the sea bottom feeding on molluscs, small fish, molluscs, worms and crustaceans. From the blue spotted lagoon stingray (the most common on reef) to the cowtail ray, bull ray and the gigantic ocean manta ray, there are 35 species known to visit the Great Barrier Reef.

Stingrays aren't aggressive towards humans, but can be dangerous if provoked or feel threatened. The well-known death of environmentalist and eco-warrior Steve Irwin has given these creatures a bad reputation. Most of the time the only reason they attack or injure a person is because we don't look out for them and stand on them unexpectedly. Relatively harmless and timid, once you get to know them they are often a delight; many scuba divers enjoy 'bucketlist' moments where stingrays take food from their hands or enjoy a belly scratch or nose rub.

However, their defences are not to be underestimated. Most of the stingrays have one or more barbed stinging spines, covered by a thin layer of skin and bacterial mucus. If threatened, the stingray will thrust its tail upwards or whip it around, lashing out, causing nasty lacerations and releasing a potent venom. The spine can break off and get embedded under the skin. Some stinging spines contain small barbs along the edges which makes removal both painful and difficult.

The stingrays - in species terms - make up almost half the wider ray family. This group includes a wide and varied array of creatures such as sawfish, skates, shovel-nose rays, eagle rays, banjo sharks (fiddler rays), guitarfish, devil rays and whip rays. They range from just a few centimetres long to the 7-metre long manta rays. The sting in the tail is what makes the stingrays different to their cousins. Skates and their relations rely on thorny projections on their skin, backs and tails for protection.

You might have guessed that the rays - including stingrays - are a sister group to the sharks, and people often mistake some ray species for sharks as their skin and tails are quite similar. The difference lies in their pectoral 'wings', often diamond, kite-like or disc-shaped with a variety of different tail adaptions.

There are two large rays opposite that both look like manta rays. One is in fact a manta ray (left) but the other is a devil ray. Adult oceanic manta rays can grow to 7 metres long, have a huge but graceful wingspan and weigh up to two tonnes. Devil rays reach only two metres, are faster and more agile and weigh in at about 300 kilograms. The mouth shape and the gills on the underbelly are also very different. However one interesting similarity is that both the manta and devil rays lack barbs on their tails: although part of the stingray family, the manta ray species has lost it barb over its evolutionary history.

THE PANCAKE SQUAD
Opposite: 150 million years of evolution has created a diverse group of hundreds of ray species. They can be skittish, timid, friendly or playful, but they come in many shapes and sizes.

QUEENSLAND SAWFISH
Sawfish are also in the larger shark and ray family. They can grow up to 7 metres long, including their rostrum (the 'saw') which is lined with teeth. Sawfish use their saw to stun or catch their prey - fish or crabs. Four species can be found in Queensland and all are listed as either Critically Endangered or Endangered, making them the most vulnerable and threatened family of all the sharks and rays. This is one of reasons the Great Barrier Reef is so important: its rivers and coastline are home to some of the last populations of these placid and elusive creatures.

A MULTITUDE OF MORAY EELS

Moray eels are ambush predators; they hide out and tend to shy away from divers. Whilst they might not be your most likely encounter when visiting the Great Barrier Reef, they would certainly one of the most fascinating. There are an estimated 200 species of moray eels worldwide and over 30 live on the Great Barrier Reef. While moray eels evoke the tales of dangerous sea serpents that's really just a myth. The majority of moray species usually only reach a metre in length, but the slender giant moray eel has been recorded as up to 4 metres long.

LET'S GET AQUINTED WITH THE NEIGHBOURHOOD
From top to bottom: The green moray can grow to 2.5 metres long, while the white-eyed moray tends to stay a little smaller. The soft fawn complexion of the grey moray, with its slender features and narrow snout or mouth, makes it stand out from the crowd. Below this is the smaller beautifully painted or adorned snow-flake moray with bright yellow spots vibrant on a black-and-white pattern that is striking to see. The ribbon moray comes in either a velvet black with flashings of white or, as pictured here, in a brilliant royal blue and yellow: its narrow body, ribbon-like movement and gulping jaws make this a treat to discover. Behind this is a zebra moray and below that is the ornately patterned dragon moray. The black and white honeycomb or tessellate moray grows quite large, but smaller than the giant moray (pictured opposite), while the yellow-eyed moray is similar to the white-eyed moray second from the top.

Moray eels' main characteristic - other than being long and thin - is that they lack pelvic and pectoral fins. Instead they have a long dorsal and ventral fin that runs the full length of the body. They are, however, still very agile hunters and move like sea snakes. When they move that is: most of the time they are holed up in dens and deep hollows scattered throughout reef territories.

The standard life span of a moray eel is around 6 to 36 years, depending on species. Starting out as thin leaf-shaped eggs then strange leptocephalus ('thin-headed') larvae, they float in the open ocean for around eight months. The juveniles, or elvers, then swim down to begin their life on the reef. Groupers, barracudas and sea snakes are among the few known predators of the elvers, but once they reach adult size there are no known predators, making many morays the apex predators in their ecosystems.

While some morays have small conical teeth, others have rows of sharp razor-like teeth, specialised for gripping their slippery prey which they then swallow whole. A few species have come to prefer crustaceans to fish: these morays hunt reef or coral crabs and have smaller, conical teeth to crush their armoured prey.

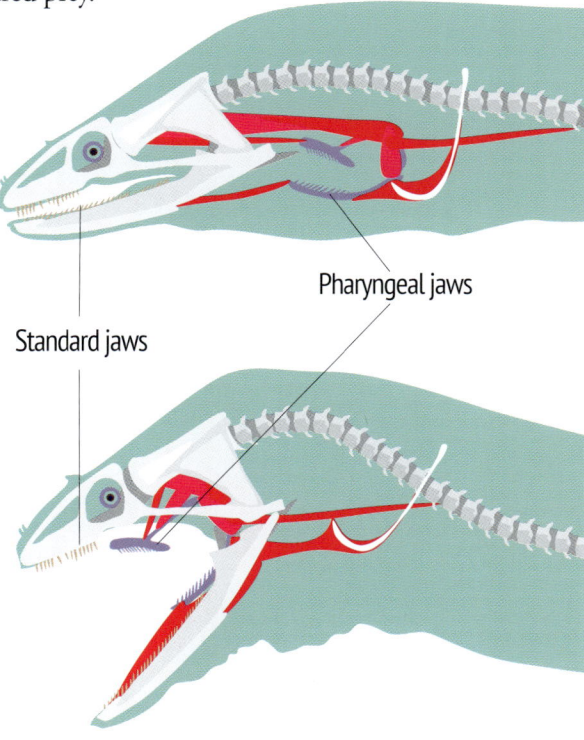

THE INFAMOUS JAWS OF THE MORAY EEL
Moray eels' extraordinary muscles attach to a second set of jaws that help them catch and swallow large prey. Like bungee-jumping cords, these muscles have the ability to slingshot their secondary pharyngeal jaws forward and backward. With sharp curved teeth, morays can catch and hold large prey such as fish, octopus, squids and crabs. These jaws inspired the double jawed xenomorph species in the popular Alien film franchise.

JELLYFISH: THE FIRST SEAFARERS

Jellyfish are truly ancient seafarers, their presence on earth dating back at least 500 million years; they are contenders for the very first creatures to swim, in the Ediacaran period. There are over 100 species of jellyfish living on the Great Barrier Reef and they come in many different shapes, colours and sizes. Jellyfish are pretty simple creatures, owing their success to the facts that they spawn in huge numbers and aren't that appealing to many creatures except turtles. Their biological design is generally fairly simple but they do have some pretty interesting adaptions: some jellyfish glow in the dark and most have tentacles with a sting to subdue their prey. Quite a few jellyfish living on the reef are quite dangerous to humans, inflicting serious, even lethal wounds.

So what exactly is a jellyfish? It's definitely not a fish, and without a backbone it's an invertebrate. Actually, it doesn't have much of anything. Jellyfish are about 95% water. They don't have a brain, a heart or blood. Their most complex systems are their nervous systems and their simple digestive systems for eating and expelling. Most are quite transparent and highly camouflaged in the ocean.

When scuba diving or snorkeling, jellyfish always seem to surprise. Whether in large groups, as far as the eye can see, or hunting the reef alone, these alien creatures instill a strange fear in many people. Most jellyfish are completely harmless to humans, but this fear isn't illogical as in Australia, Irukandji, bluebottles (also know as the Portuguese man o' war) and box jellyfish can be very dangerous to swimmers (note that bluebottles, though very jellyfish-like, are actually colonies of four different simple animal species).

While bluebottle stings aren't fatal in most cases, they do pack a nasty punch, leaving skin red, swollen, blistered and sometimes scarred. These small organisms float on the surface, using their bodies as sails to move through the water, and so can be avoided. Box jellyfish come in over 30 different species with *Chironex fleckeri* and *Carukia barnesi* being the most potent. Stings from these, and a few other species, are very painful and can be fatal. Box jellyfish are a generally little larger than bluebottles, and are again usually visible in the water. However, our final subject, with an infamous 'get out of the water' reputation, is one of the smallest, hard to see, yet most venomous jellyfish in the world.

There are over a dozen species of Irukandji and they are closely related to the box jellyfish. Only around one cubic centimetre when fully grown (plus up to a metre of tentacles), their diminutive size simply increases the danger they pose. They live in tidal streams across the Coral Sea and throughout the Great Barrier Reef year-round, but appear in greater numbers off Queensland beaches around the 'jellyfish season' spawning time (November to May).

Jellyfish reproduction is quite complex; it varies slightly between species, but the cycle is common across most. The adults (the 'medusa') mate to produce fertilised eggs, which hatch into planula larvae. After leaving its parent, the larva swims away and attaches to the reef, where it grows into an anemone-like polyp before severing its tether and swimming off as a new fledgling medusa (called an ephyra). Polyps can also reproduce by budding asexually. The adults are either male or female, yet some species change sex as they age. This is not uncommon in many different marine species.

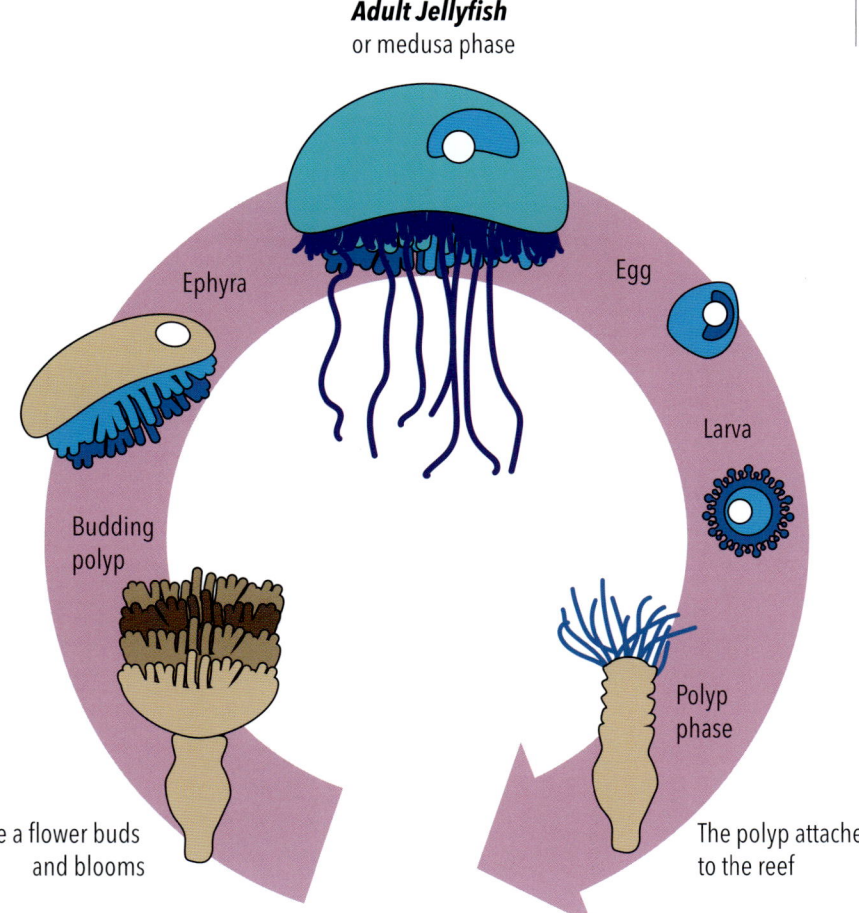

Adult Jellyfish or medusa phase
Ephyra
Egg
Larva
Polyp phase
Budding polyp
Like a flower buds and blooms
The polyp attaches to the reef

JELLYFISH COME IN MANY SHAPES AND SIZES
Opposite: Floating on the surface, bluebottles use their bodies as sails. While some jellyfish have relatively small solid arms, others have a long tangle of dangling tentacles. Some jellyfish are almost transparent, others quite opaque. They vary in size from only a cubic centimetre to 1-2 metres across (with tentacles tens of metres long). Most are actually quite harmless.

THE GREATEST OCTOPUSES' GARDEN

As the greatest living structure on planet earth, the Great Barrier Reef is surely the greatest octopuses' garden in the world! Cephalopods (octopuses, squids and their kin) have been around a very long time, splitting from their sister group the gastropods (snails etc) nearly half a billion years ago. So what is a cephalopod? It's generally a big-eyed, multi-armed, often shape-shifting, venomous mollusc with nine 'brains' that can arguably match our wits and intelligence. The best known of the family are the shell-less, eight-limbed Octopoda, which includes about 300 species worldwide. Their close cousins are the squids, cuttlefish, nautiloids and the sadly extinct but once dominant ammonites. The octopuses are particularly strange creatures, with three hearts, blue blood, a mastery of camouflage and ink jet propulsion for protection. All octopuses (plus all cuttlefishes and a few squids) are venomous, although only the blue-ring is potentially dangerous to humans.

Well if those facts aren't weird and wonderful enough, the story contines!

An octopus has four stages in its life cycle - egg, larva, juvenile, and adult. The male is often a great deal smaller than the female and in some cases they look like a completely different species. This is not the only odd thing about these creatures. Both male and female die within a few months of first mating, never mating again. The female lives long enough to brood and hatch to her nest of eggs. It's an unusual reproductive strategy (known as semelparity), although not unique to them, with most octopuses living only 1 to 5 years. Squids and cuttlefish are also semelparous, the nautilus being the exception in the family.

Yet another other interesting fact about the female octopus is that different species sit on their eggs for very different periods of time. The common octopus may take around 50 days to nurse her eggs in the warm shallow waters of the Great Barrier Reef; the incubation period for the giant Pacific octopus, living in cooler water, can take around 6-7 months. The longest incubation period was a recently observed deep sea octopus which took 53 months years to incubate her eggs (and die straight after) - this is the longest known egg incubation in the animal kingdom.

Yes, selfless mothers and dedicated homemakers, but with a suicidal death wish. After mating with (and often killing) the male octopus, the mother octopus retreats to her den. She expels each egg, one by one, over several days if not weeks: up to 50,000 of them. She braids them together into long chains and attaches them to the roof of her cave. Although in a sort of suspended animation, the mother octopus then cares for and protects her eggs continuously, never eating, until they hatch and she dies.

Over the course of the incubation she continuously pumps oxygenated water over the eggs. During this time her body breaks down, tissues and organs deteriorate, reducing her movement and often causing white, unhealing lesions to form on her body. The octopus hatchlings are born unaware of their mother's care - without the guidance of parents they immediately become reef hunters, seeking out food and living by their fast-forming wits. Only a fraction of 1% will survive to reproduce themselves.

That's just the octopus! Cuttlefish and squid have similar life cycles, mating once and dying soon after. Surprisingly little is known about the nautilus. In Australia cuttlefish come together in groups and face-off for a female's attention after battling it out. She lays her eggs in crevices on the reef. They need to be protected and hidden since she will die at the end of the breeding season.

Octopuses in particular make for wonderful meals, so have become masters of disguise and camouflage; changing colour, mimicking coral shapes and textures. They also have their amazing ink. When in danger they squirt it behind them, instantly obscuring a predator's view as they zoom away. But it's not just a smoke screen: the ink contains tyrosinase, a compound that burns predators' eyes and temporarily paralyzes both their sense of smell and taste, something like a pepper-spray!

IN THE DEEPEST RECESSESS OF THE GREAT BARRIER REEF

What must be the cutest octopus in the world, the dumbo octopus, was filmed in 2020 by a science team researching the inky depths of the Coral Sea adjacent to the Great Barrier Reef. Down at depths of over 620 metres, the expedition also mapped more than 13,000 square miles of seabed and discovered an abundance of nautili.

THE ALIEN UNDERSEA OCTOPUS WORLD

The Great Barrier Reef is home to over a dozen known species of octopus. Yet, they are so good at camouflage and knowing how to avoid being seen, that there could be many more to be discovered. Four different new species of ocellate octopuses were discovered in one year in 1992.

THE WEIRD AND WILD WORLD OF NUDIBRANCHS
The Spanish dancer is one of the most iconic of all the nudibranchs (pictured floating), yet below this beautiful display are many of the other nudibranchs you may come across throughout the reef. There are so many species it is hard to identify them all. Some have fun nicknames, such as "sea goddess", "splendid", "dragon" and "painted". Can you spot the flat worm hiding amongst the nudibranchs?

THE SECRET WORLD OF NUDIBRANCHS

You'd think with all that colour and and vibrancy that these jewels of the marine world would stand out, but they are actually quite hard to spot until you really know what you are looking for. Nudibranchs (pronounced "nudeebranks") range in size from a few millimetres in length to over 50 centimetres, but most of the 1000-odd different species on the Great Barrier Reef are between 1.5 and 4 centimetres long as adults. Over 300 species have been recorded in the reefs surrounding Heron Island alone and there are probably many more still hiding undiscovered in the coral. Just like the slugs in our garden, these nocturnal creatures mostly remain tucked away. The bright colours may ward off some predators, but they also make for great camouflage amongst the coral. Herbivores, they graze on algae, helping keep the reef clean. The name nudibranch means 'naked gills' and the difference between these sea slugs and their close kin is that they breathe via exposed tufts of gills, normally on the rear of their bodies.

Nudibranchs account for about a third of over 3000 known molluscs living on the Great Barrier Reef. Their cousins include clams, oysters, other shellfish, squid, octopus, cuttlefish, nautilus, chitons, snails, and sea slugs. Molluscs play an integral role on the reef, providing a source of food for many of the other reef creatures and in turn doing their part to keep the reef healthy.

Nudibranchs are soft bodied animals, but as larvae or in the egg, some have a hard shell for protection, which they shed as they mature. They are either multi-coloured or brightly coloured and come in a huge range of shapes and sizes, ranging from a simple slug shape to highly complex body plans. Some, such as the blue dragons, use poisonous chemicals to deter predators, and many nudibranchs sport a red coat to let predators know they aren't good to eat.

The most famous nudibranch is the beautiful Spanish dancer. These, the largest nudibranchs (with reports of animals up to 60 centimetres long), are often spotted by divers swimming between coral heads in a dazzling display. Their bodies rhythmically ripple and flow as they swim, just like the ornate skirts of a Spanish flamenco dancer.

Along with the nudibranchs' unique feathery gills on their backs, through which they breathe, a key characteristic is the two horns on their head known as rhinophores. Nudibranchs have tiny eyes and pretty useless eyesight, so they use these receptors to pick up traces of chemicals in the water, helping then find food and navigate their environment.

While most nudibranch species are concentrated in tropical waters such as our Great Barrier Reef, they have been found all over our oceans and at great depths too. Like snails in our garden, these gastropods do leave a slime trail. The female can lay up to two million eggs at a time! These egg masses are laid in either spirals or coils of jelly on the reef.

My first encounter with a nudibranch was when diving on a reef in the Bass Strait. Since then I have discovered them in all sorts of places. A friend of mine actually photographs and catalogues them. He started out attempting to record 1000 nudibranchs before he died and in the end found close to 3000.

THE BLUE DRAGONS
The nomadic blue dragons feed on both blue bottles and blue button jellyfish (pictured right). The dragon can ingest and store the stinging particles from their prey and use them to deliver a powerful sting themselves.

THE SHELLY KING

There are many varieties of large clams, but the largest bivalve mollusc is the giant clam. This colossal creature weighs, on average, around 150 kilograms but can be much more hefty. But what truly draws the attention is the wide variety of iridescent colours that seem to glow from their large gapping lips and soft fleshy structure. Created by algae inside the clam, together with the clam's natural pigment, this vibrancy indicates the clam's health. Dying clams turn white, their skin bleached. Healthy giant clams can live up to 100 years, grow well over a metre across and weigh up to 300 kilograms. Hiding in the reef around this clam are two strange sea cucumbers. The purple elephant trunk sea cucumber is a delicacy in Japan.

MORE MOLLUSCS: THE WILD WORLD OF CLAMS, OYSTERS SNAILS AND OTHERS

Australia can't claim to have the greatest range of clams. We do have a small range of mussels, surf clams, Venus shells, Sydney cockles, blood cockles, scallops, razor clams and wild oysters: all bivalved molluscs. We do have our fair share of marine snails, sea slugs and their kin, including families such as the conchs, periwinkles, abalone, turbos, conical limpets, whelks, the highly venomous cone shells and of course the already-met nudibranchs. You will certainly find the bleached remains of the shelled varieties beachcombing the shores across the Great Barrier Reef, but it's far more satisfying seeing these creatures in action amongst the coral or popping out of the sand. Here's a secret: wild oysters, plucked straight from the rocks and eaten raw, salted by sea water, are the most authentic way to experience their flavour (if shellfish is on your menu).

There are over 3,000 known molluscs on the Great Barrier Reef. Shellfish (bivalves and others) and gastropods (including nudibranchs) account for the majority of these and cephalopods make up the rest. The whole huge family (more species than any other phylum in the oceans) are obviously delicious. Sea slugs, sea snails, octopuses, mussels, oysters etc. are a huge food source for sea birds, fish, mammals and other marine creatures on the reef, and humans foraged for them long before our ancestors had tools to hunt with.

Fish have developed tusk like teeth and unique strategies to crack open molluscan shells and get the mouthwatering treasures inside. Sea birds like oyster catchers specialise in clams and herons will not miss an opportunity to grab a mollusc when stalking the tidal pools (which play a huge role in the food chain). Some sea snails even eat other sea snails. You can tell the difference because carnivorous snails are shaped like a gravy server, are spearlike or coned while the algae eaters and herbivores are are spiral-like (conical, triangular or rounded) with larger bowl openings where the foot of the snail comes out to graze the coral. You can tell if a snail was killed by another snail by a distinctive hole drilled into the shell.

THE TIDAL POOLS

Below: all along the coast where the rainforest meets the reef, there are many tidal pools and fringe reefs. In these waters filter-feeding mussels and oysters grow in abundance. But lurking amongst the coral, anemones and graveyard of shells is the infamous painted cone shell! Hunting even small fish this carnivorous snail has a venomous stinger: it fires a toxic dart that can put you in hospital if you aren't careful!

COLOURFUL CRUSTACEANS

It is estimated that roughly 1300 species of the huge Crustacea family have been found on the Great Barrier Reef, but the actual total living there could be far, far higher as many are microscopic and many others are highly camouflaged and very good at remaining unseen. This diverse group includes krill, lobsters and crayfish, all the various shrimps and crabs, and the family that Charles Darwin was instrumental in first describing, the barnacles.

Crustaceans are fascinating for many reasons. Take the decorator crab, which promotes the growth of coral on it's carapace to camouflage itself and make it unattractive for prey. Or the painted lobster which has some of the best camouflage on the reef, with hypnotic confusing patterns to blend in with the coral. Barnacles and anemone crabs have fine hairs on their appendages to filter food. Giant, kaleidoscopically coloured tropical lobsters march from reef to reef. The territorial giant mantis shrimps and their deadly martial arts skills: their stunningly fast forelimbs dispatch their prey in an almost unbelievable display of underwater Kung fu. Soldier crabs, mud crabs, Moreton bay bugs and sand crabs burrow swiftly into the sand to disappear at any sign of trouble. Blue swimmer crabs swimming away from danger. Actually, most crabs swim well, but most will first choose rock crevices and other hidey holes as their preferred evasive manoeuvre.

IDENTIFYING CRUSTACEANS
Right, top row: decorator crab, soldier crab, honeycomb coral crab, red reef crab, hermit crab.
Row two: a prawn and a banded coral shrimp.
Row three: mantis shrimp.
Row four: a Moreton Bay bug, also known as a mudbug, Northern Bay lobster, sandbug, shovel-nosed lobster, slipper lobster or simply as a bug.
Row five: blue swimmer crab and a sand crab.
Row six: green mud crab.
Row seven: painted lobster (or tropical rock-lobster), similar to the southern red rock-lobster except their carapace or shell is ornately coloured. Both the painted and red lobsters reside on the Great Barrier Reef.

CRABS OF MUD, SHELLS AND SAND
Below: demonstrating it's foreboding power, the green mud crab defends itself on the mangrove mud flats of estuaries and tidal flats. Prized for its delicious meat it's a unfortunately a hopeless effort when facing those who know how to avoid their huge nippers. In the background millions of soldier crabs emerge from the sand and are on the march. The unconcerned, lone hermit crab hauls his home (an acquired empty shell) across the mudflat to a tidal pool to beat the heat of the midday sun.

ARMOUR AS ORNATE AS A SAMURAI
The gorgeously painted tropical lobster hides out in a cubby-like coral hollow by day and hunts at night. This is great camouflage, but others have better: can you see the many other crustaceans in this picture?

OUR BELOVED SEA TURTLES

The Great Barrier Reef is a protected sanctuary for six of the world's seven known species of sea turtles. While some species live in the waters around Australia, others are rarely seen and migrate great distances across the globe: they are known as the 'ancient mariners' of the sea. Turtles have been navigating our oceans for more than 150 million years and have changed little over that time. Much of what we know about turtles comes from 30 years of research conducted by the Queensland Turtle Conservation project and from existing knowledge. All of these turtles are listed as endangered to some degree, loggerheads and green turtles are more common and often seen on the reef, while the olive ridley and the critically endangered leatherback turtles, though known to frequent the reef, are rarely seen. The critically endangered hawksbill is another turtle that utilises our shores and reefs. The flatback is endemic to the coastal waters of the continental shelf between Australia and Papua New Guinea, with breeding sites at suitable sandy beaches right down the reef.

All marine turtles have a similar life cycle. Those young turtles that survive the skitter down the beach and the dangerous first few weeks adrift on the ocean currents for many years then find sanctuary in the reefs, dieting on jellyfish or, in the case of the hawksbill, coral sponges. They're slow growing, taking decades to reach maturity (20-50 years depending on the species). When they mature they start their migration from the relative safety of their feeding grounds to their ancestral nesting sites.

Mating generally takes place offshore, and most turtle species mate with a number of different partners. The vigorous competition can often exhaust the female turtles: eager males can sometimes drown or beach them on their backs. The females store the sperm to fertilise several clutches of eggs which they lay in roughly fortnightly trips to the beach across the summer nesting season.

Each nest contains around 120 eggs, ranging in size depending on the species. After laying between three and seven clutches across the season, the female crawls back to the sea exhausted and may not return to nest again for up to eight years.

After about seven to twelve weeks, if the nests aren't predated or disturbed, the baby turtles emerge, digging their way out onto the beach and making a mad dash for ocean. Crossing the beach is very dangerous for these newborns. Even so, they immediately imprint themselves with the necessary cues to find their way back here when it's their time to breed.

Feeding on tiny sea animals, the hatchlings swim out to the open ocean. Hiding on floating seaweed mats caught up in ocean currents to avoid birds, here they remain for up to ten years: not much is known about this part of their lives. The adolescent turtles migrate back to begin their inshore foraging amongst the reef and island areas. They remain in these sanctuaries until they are sexually mature and the cycle begins again.

SEA TURTLE BREEDING

● **Green turtles** are found the world over and they are the most abundant of all the species of turtle found on the reef, in two distinct populations: one in the north and one in the south. They eat algae, seagrass, mangrove fruit and jellyfish. They migrate to Indonesia, the Gulf of Carpentaria, Arnhem Land, the Torres Strait, Papua New Guinea, the Solomon Islands, Vanuatu and New Caledonia. ● **Loggerheads** feed on animals such as crabs, sea urchins and jellyfish. Two genetic groups inhabit Australia, one on the Great Barrier Reef the other on the west coast. They have a small range of migration to the Gulf of Carpentaria, Arnhem Land, Torres Strait, and Papua New Guinea. ● **Hawksbill** nesting is sadly in decline on the reef. They migrate to the Solomon Islands, Vanuatu and New Caledonia to forage on coral sponges. ● **Flatbacks** nest in the southern part of the reef and their local migration to the Torres Strait and the Gulf of Carpentaria makes them endemic to the area. The ● **olive ridley** is the smallest of the Australian sea turtles, nesting in the remote north. The ● **leatherbacks** are the largest species, with soft skin and shells with unique, distinguishable ridges. They are elusive reef visitors, sometimes sighted on migrations south and nesting rarely on our shores.

TURTLE BEACHES
While Australian flatback turtles like clean sandy beaches and use both the mainland and the islands in the southern reef region, green turtles (pictured here) favour the islands to protect their eggs from predators.

TURTLES OF THE GREAT BARRIER REEF
Some remoras ('sucker' fish) swamp the soft skin of a leatherback turtle, hitching a ride in the open ocean. Catching the warm rays of sunrise a flatback takes a dive to hunt after taking a breath at the surface. Behind the flatback is a loggerhead turtle, also diving after surfacing. A green turtle floats in the current, some remoras in his slipstream; several others dash away, startled by the leatherbacks. An olive ridley turtle sneaks past a graceful leatherback in the depths. Rising to the surface is a dappled-faced hawksbill.

AN UNDERWORLD OF OUR BEAUTIFUL SEA SNAKES

Australia has around 140 species of land snakes and 32 species of sea snakes. There are sea snake populations right around the northern seas lapping Western Australia, the Northern Territory and Queensland. Around the world, there are nearly 60 species of sea snakes and, partly because the Great Barrier Reef borders the Coral Triangle to the north, about a quarter of these are resident on the reef. It's believed that sea snakes evolved from elapid (front fanged) land snakes about 30 million years ago, but a great deal is still a mystery when it comes sea snakes.

Sea snakes are some of the most venomous creatures on the planet, but lucky for us they are pretty placid and curious creatures. Most bites to humans are a result of the sea snake feeling threatened, not due to the snake being aggressive. So handling or touching these creatures in the water is ill advised. Most land snakes are actually well adapted swimmers and our indigenous traditional custodians relate snakes to water spirits and to dreamtime stories of rain and the formation of river courses. So it's not hard to imagine the evolution of the sea snake. It's thought that the sea snakes evolved from tiger snake relatives and took to the shallow fringe reefs and mangrove ecosystems to hunt for fish and crustaceans. They adapted to their salty environment by evolving a paddled tail, skin to allow them long exposure to sea water and the skill to free dive for long periods.

Most sea snakes are solitary, and they need their highly toxic venom to dispatch their prey quickly and efficiently. There is an exception: sea kraits hunt in packs. Some sea kraits also differ from their cousins in another critical way. Most sea snakes gestate their eggs within their bodies, giving birth to live young. The banded sea kraits, however, come to shore to lay their eggs. Everything else, though, including courtship and mating is done beneath the waves.

In 2016, marine ecologists working in Cleveland Bay, off Townsville, made the discovery of a species not seen on the reef before: the spine-bellied sea snake. Yes, in the heart of the Great Barrier Reef these beautiful sea snakes were discovered right under the noses of the researchers. And it wasn't just another sea snake to add to the growing list of Great Barrier Reef sea snakes, but a reef nursery full of spine-bellied sea snake juveniles and pregnant females. This is the first sea snake nursery ever discovered.

A TANGLED SEA OF SNAKES

Opposite: There are well over a dozen species of sea snake to be found on the Great Barrier Reef. Here are 17 sea snakes you could potentially come across on the reef. While some are rare and are more elusive than others, all produce lethal venom to paralyse their prey. Yet they are rarely aggressive towards snorkelers.

THE PACK HUNTING SEA KRAIT

While most sea snakes are solitary by nature, the curious and highly venomous sea krait is a pack hunter. I've personally witnessed these slithering wolves of the sea hunting coral heads in large numbers, seeking and stalking every nook and cranny. Working together, these lethal predators hunt out small crustaceans, octopus eggs and tiny fish.

THE SALTWATER CROCODILE
The remote beaches and waterways of the northern Great Barrier Reef, north of Proserpine and the Whitsundays, are patrolled not just by bull sharks, but by the infamous 'salties' or saltwater crocodiles. Even in crystal clear blue water these ancient leviathans can remain undetected, awaiting the chance to ambush their prey.

LAND REPTILES OF THE REEF

Turtles and sea snakes aren't the only reptiles on the Great Barrier Reef. Plenty of other, more terrestrial reptiles are common on the reef islands off the coast. The biggest and most feared is the saltwater crocodile, but there are plenty of others that can be very dangerous if provoked, such as tiger snakes, eastern brown snakes, death adders and coastal taipans. When you visit Whitehaven Beach it's not unusual to be greeted by goannas (lace monitors) wandering the picnic tables looking for food scraps left by careless tourists.

If you look carefully and take your time while hiking to the summits of the islands off the coast of Queensland, you may hear a rustle in the leaf litter and discover a large number of reptiles living in the well established bush that adorns the hillsides.

With plenty of food (mostly nesting birds, frogs and rodents), goannas, snakes and many skinks and geckos have survived on these islands since their separation from the mainland a few thousand years ago. Common skinks make up the bulk of the reptiles, but there are many more to look out for.

Whitsunday Island alone is home to over 25 reptile species including the elusive Boyd's forest dragon, mangrove and water monitors, various snakes, the gorgeously painted blue-throated skink and the wood gecko, just to name a few.

New species are discovered all the time. In 2023, on a remote, uninhabited, boulder-strewn island 50 kilometres offshore from Mackay, in the southern reaches of the reef, a large leaf-tailed gecko with an interesting 'beak-like' face, skinny legs and a spiky tail was found. Remaining hidden throughout the day, the Scawfell Island leaf-tailed gecko beats the heat then comes out at night to hunt. With finely dappled skin colouring, it is perfectly adapted to remain camouflaged on the gigantic boulders that make up much of the island's habitat. Finding such a large (at 10 centimetres) gecko is both surprising and exciting. There is an urgency to learn about this species as the population size could be as few as 30 individuals, but for now scientists are overjoyed with their new find.

ISLAND SURVIVORS
Below: the saltwater crocodile (in the background), forest dragon, water dragon, lace monitor, the slender, yellow-bellied water monitor, the newly discovered Scawfell Island leaf-tailed gecko, the wood gecko and the blue-throated skink all call the Great Barrier Reef home.

BIRDS OF THE REEF

The Great Barrier Reef regularly supports around 200 different species of birds, from coastal shore birds to the ocean-hunting sea birds, and if you include all species that have ever been sighted on or close to the reef, the number increases to an astonishing 500 or more. There are island nesters, ocean wanderers, wetland waders, mangrove stalkers and mudflat sifters and pipers. Also present are the long-distance migrating aviators, soaring birds of prey, rainforest birds and bower birds. On the coast and larger islands are also the colourful parrots, nectar collectors and tropical honeyeaters.

Birds not only nest and feed on the reef, they are an important part of the ecosystem. Just one important connection between birds and the undersea world is ... their poop! Guano is the accumulated excrement of seabirds or bats. Containing a high content of nitrogen, phosphate and potassium from their diets, rainfall and coastal storms erode and wash this precious fertiliser onto the reef. These are key nutrients for coral growth, they balance pH levels in the water and contribute to algal blooms which provide food for many fish colonies on the reef. Most nesting coral cays have thriving reefs fringing them; small coral fish shelter in these coral gardens and larger fish use these grottoes as nurseries for their young.

Ocean roaming shearwaters and jet black Parasitic Jaegers forage far out at sea and return to the reef to deposit their waste, which in turn helps the reef nurse the fish they eat. Coastal birds like terns and seagulls hunt in the the tidal pools and leave their droppings on the exposed parts of the reef. High tides then wash the nutrients back into the water. Bird poop is so important for the cycle of life on the reef.

When we we think of birds on the reef, we immediately think of seabirds. But the coral cays are not just sanctuaries for ocean birds; amongst many others, White-bellied Sea Eagles and Ospreys create massive nests of sticks and branches. These nests are maintained over decades, with sea eagles using them for many seasons in succession, and they get to huge sizes. They can be nearly three metres across and just as high as each year more material is added. While most nests are located in trees, high above the ground, some are perched on cliffs or rocky outcrops.

COASTAL BIRDS
Above: the Olive-backed Sunbird and the Blue-faced Honeyeater are two of the beautiful birds you might spot on the coast taking advantage of tropical flowers. **Below:** the beaches, mangroves, islands and rich estuaries along the Queensland coast host many shore birds; terns, seagulls, oystercatchers, stilts, Red-necked Avocets, Eastern Curlews as well as migrating knots and sandpipers. **Opposite:** islands are great safe havens and nesting sites for seafarers such as shearwaters, Parasitic Jeagars, frigatebirds, crested terns and Silver Gulls.

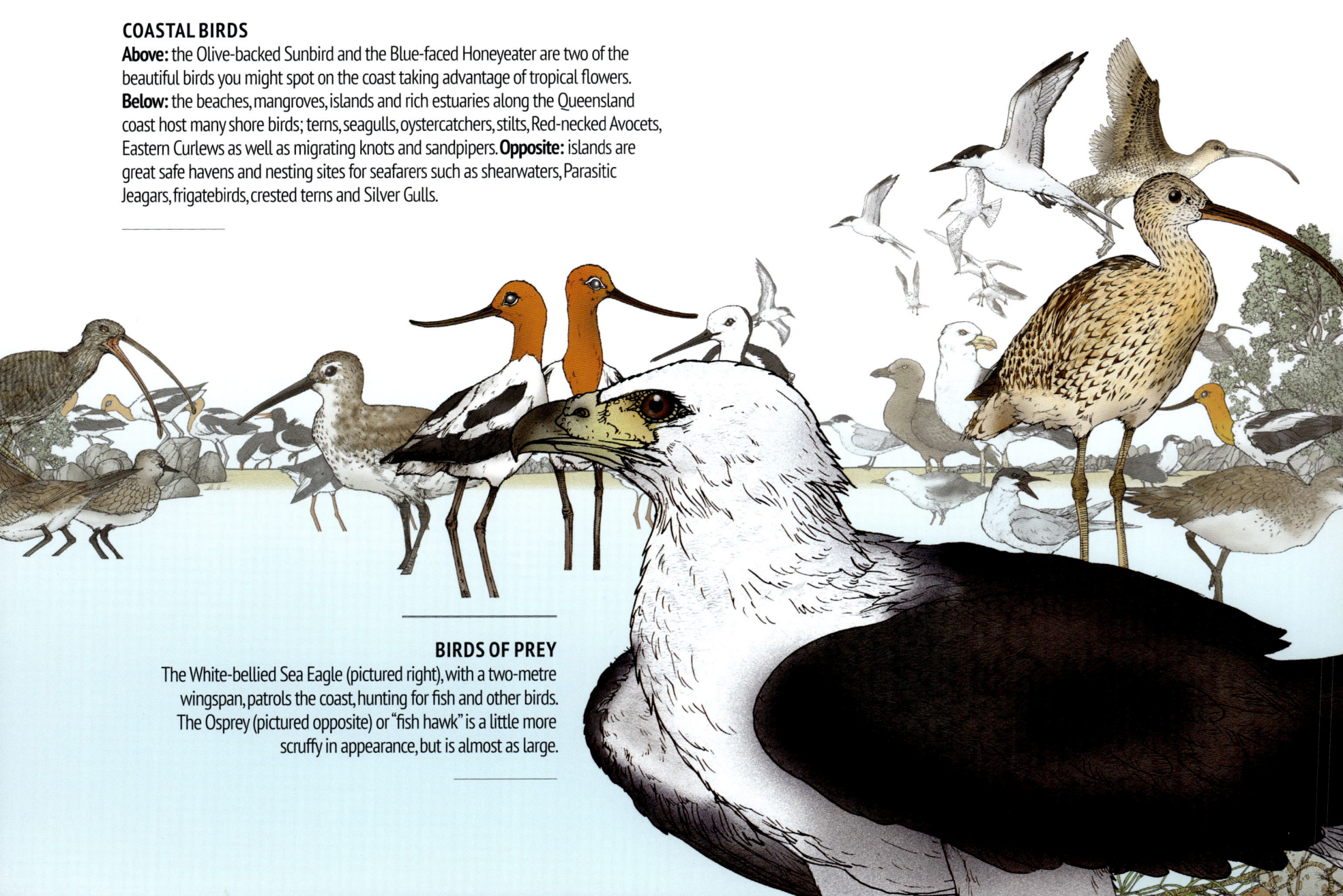

BIRDS OF PREY
The White-bellied Sea Eagle (pictured right), with a two-metre wingspan, patrols the coast, hunting for fish and other birds. The Osprey (pictured opposite) or "fish hawk" is a little more scruffy in appearance, but is almost as large.

MIGRATING GIANTS AND RESIDENT DOLPHINS

It is quite awe-inspiring watching, from the edge of your boat, a 16-metre, 25-tonne humpback whale breach and splash down in Hervey Bay. One of the largest species of baleen whale, humpbacks are known for dazzling displays that make them popular with whale watchers. These ocean giants migrate every winter to give birth and nurse their young in the relatively safe waters of the Great Barrier Reef. These playful creatures only hunt in their summer Antarctic haunts, eating krill, small crustaceans and schooling fish, living off their blubber through the winter months. They share the reef with over 30 different species of whales and dolphins.

Whales and dolphins hold a special significance for many who use and visit the Great Barrier Reef. For many it's simply whale watching but for the traditional custodians such as the Woppaburra people of the Keppel Islands, the Mugga Mugga or whale is a totem that connects them to country, both land and sea, as well as to their ancestors. The health of the whale population is important to their culture.

Dwarf minke whales, humpback whales and bottlenose dolphins are among the most commonly sighted species on the reef, yet other species are often reported including the vulnerable Indo-Pacific humpback dolphins, Australia's own snubfin dolphins, spinner dolphins, pan-tropical spotted dolphins, false killer whales, killer whales, short-finned pilot whales and Brydes whales. Even sperm whales and various beaked whales are occasionally seen, and a Longman's beaked whale was once stranded on the beach at Mackay.

Whales and dolphins are some of the ocean's most intelligent creatures. Humpback whales have brains weighing several kilograms. They migrate over 25,000 kilometres a year, remember locations, are well organised, social, and have been observed playing with and protecting other species such as dolphins, seals, smaller whales, and even on occasion humans. There is plenty of evidence, that humpbacks and dolphins, just like elephants, can recognise and remember individual humans. Most humans can barely tell one whale species from another, let alone tell two individual dolphins apart!

HAVING A WHALE OF A TIME
Opposite: humpback whales migrate to Hervey Bay and use the Great Barrier Reef every winter to calve and nurse their young before returning to Antarctica to feed in summer. They share the reef with about 30 species of whales and dolphins, including these playful bottlenose dolphins.

DOLPHINS OF THE REEF
Right: There are several types of dolphin that make use of the Great Barrier Reef's rich resources. Illustrated right are four species starting with the endemic and most unique looking snubfin, then the common bottlenose, the acrobatic spinner and the strange and vulnerable Indo-Pacific humpback dolphin.
Left: is a location map of our most threatened and vulnerable dolphins, the snubfin and humpback.

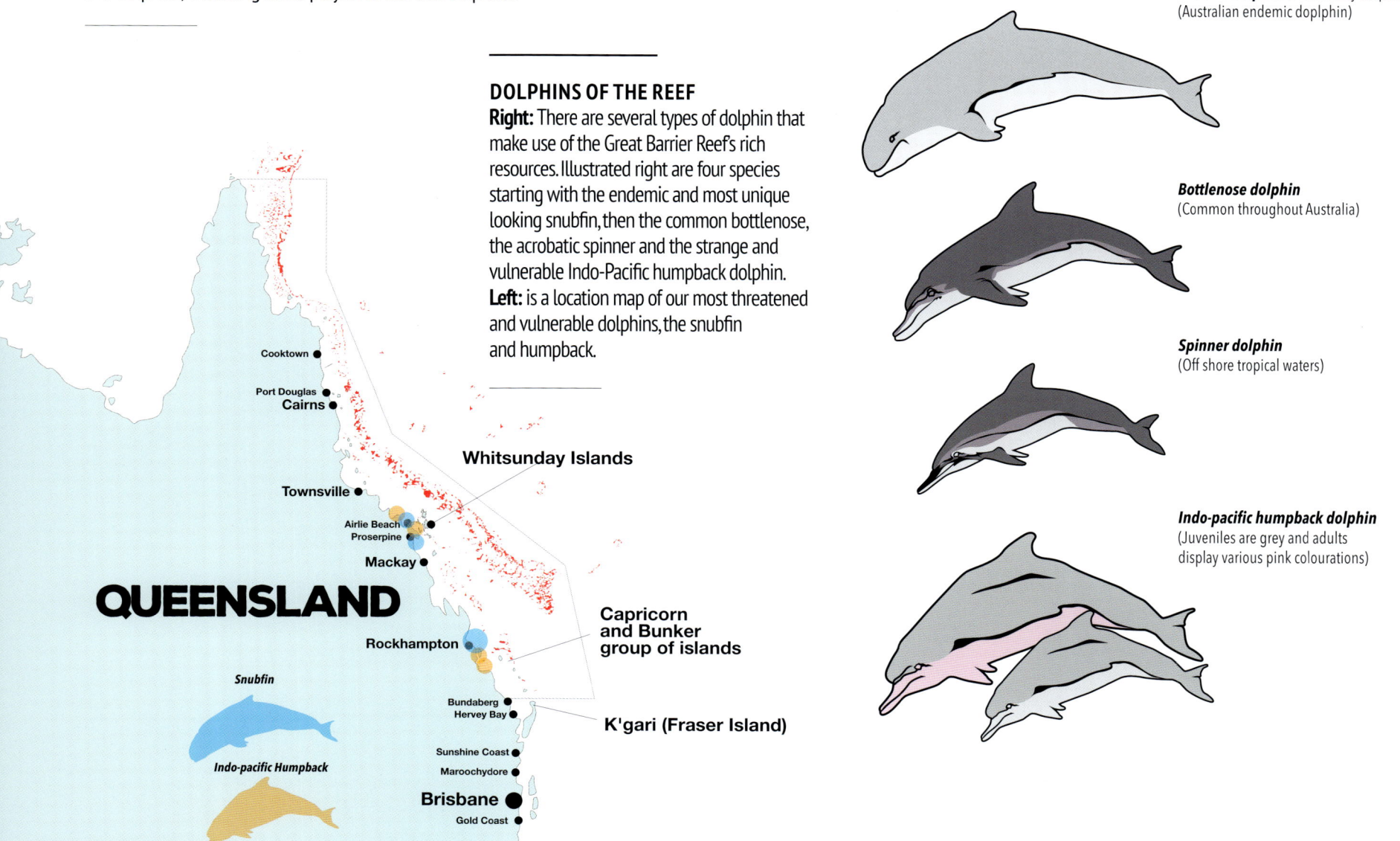

Snubfin dolphin or Irrawaddy dolphin (Australian endemic dolphin)

Bottlenose dolphin (Common throughout Australia)

Spinner dolphin (Off shore tropical waters)

Indo-pacific humpback dolphin (Juveniles are grey and adults display various pink colourations)

OTHER MAMMALS OF THE REEF

Whales and dolphins are not by any means the only mammals to be found on the Great Barrier Reef. There are a few other marine species and the more than 900 islands, dotted from the tip of Cape York all the way down to K'gari, many densely forested, plus the adjacent mainland coasts, are home to a host of terrestrial species. Fascinatingly, their mere presence holds secrets to the ancient beginnings of the reef and its modern connection to the mainland.

Right across the Great Barrier Reef, from the Cape York to Moreton Bay, you find dugongs, though their populations are patchy and considerably less than they were just 200 years ago. Feeding on seagrass, these aquatic vegetarians are now mostly found in Moreton Bay Marine Park, Amity Banks and Pumicestone Passage. Moreton Bay Marine Park's dugongs number between 600-800, perhaps 10% of the total on the reef. These fascinating creatures are commonly seen in pairs but can group together in herds of up to 100.

Although more commonly seen than dugongs, the only distantly related sea lions prefer the cooler waters south of Brisbane though they do venture further north, depending on the currents, seasons and water temperature. They have been seen hunting and frequenting the southern reef and its islands and are contributors to the reef ecosystem. The relative paucity of sea lions might be one the reason the great white shark doesn't venture north into the reef.

There are a large number of mammalian island dwellers on the Great Barrier Reef, but many are finding life tough going. At the most northern point of the reef, Bramble Cay was the scene of the first extinction of a mammal species that has been directly caused by climate change. The loss of the Bramble Cay melomys highlights the urgency to find ways to protect island-dwelling and coastal mammals. The fawn-footed melomys is a related small rodent, but with a more stable population and is found right down the coast of Queensland. Island environments demonstrate the incredible adaptability and resilience of many species, but also their vulnerability. The unadorned rock-wallaby lives on Whitsunday Island. Cut off from the mainland for several thousand years, it has evolved into a distinct species, quite different to its sister species, the Proserpine rock-wallaby. Even though it's on the mainland, the Proserpine rock-wallaby, like the unadorned rock-wallaby, has become isolated in a small area of wilderness, but surrounded by an ocean of cane fields and other agriculture instead of salt water.

Another surviving relative of these wallabies is Bennett's tree-kangaroo which inhabits the wet tropics. With its distinctive thick reddish fur, long tail, and stocky legs and arms, it still survives in the Daintree and protected rainforests of the far north - but for how long?

Unsurprisingly bats also have found a place on many of the reef's islands. The little bent-winged bats and the black flying foxes of the Whitsunday island group are good indicators of the diversity of the reef's bats. Like bird poop, the bats' guano is a significant addition to the nutrient cycle on the reef, helping trees establish a better foothold, balancing the pH of the soil, and bringing nitrates to the reef in storm run off.

THE INTERCONNECTION OF LAND AIR AND SEA
Left: while bent-winged bats and flying foxes take to the air at night, dingos on K'gari patrol the beaches, beach-combing for what the sea serves up. A fawn-footed melomys also searches for food alongside its lost cousin, the Bramble Cay melomys. Creamy-breasted unadorned rock-wallabies have been isolated by rising sea levels and time on Whitsunday Island while the Proserpine rock-wallaby is equally isolated, but by mankind's agriculture. A Bennett's tree-kangaroo hides out in the Wet Tropics. Rarely seen sea lions play with the dugongs, completing the interconnected trinity of land, air and sea mammals.

GLACIAL AND INTERGLACIAL TIMES
Right: continental islands like the Whitsunday group hold the key to the many secrets of the creation of the Great Barrier Reef. Various rock layers and their entombed fossils, plus the evolutionary heritage of its current denizens, help tell the tale of the current three million-year-old ice age. The story includes the dramatic fall and rise (during the interglacial periods) of the sea level. This has happened dozens of times over, often allowing time for the creation of reefs, islands and - almost certainly - entire suites of new animal and plant species.

Glacial Period
10,000 years ago

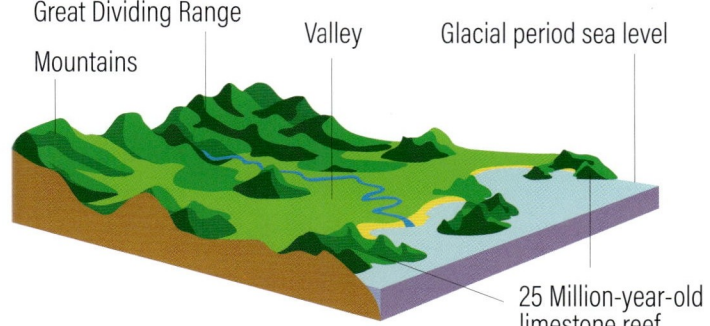

Interglacial Period
Present day sea levels

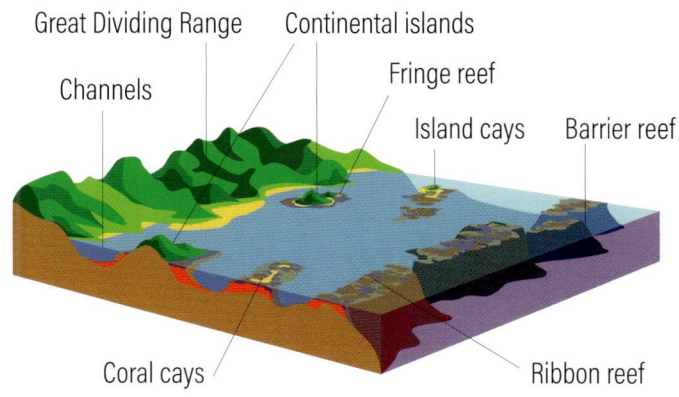

The end... the tragic story of the Great Barrier Reef's only endemic mammal, the Bramble Cay melomys

In 2016 researchers monitoring the Bramble Cay melomys, a small endemic rodent clinging to life only on the Great Barrier Reef's most northern coral cay (island), declared that it had completely disappeared. In 2019 it was declared extinct. We may have had a chance to save it but for its remote, singular location. It's a cautionary story and one we should all come to know: Bramble Cay is a tiny refuge and nesting site for sea birds and turtles. The 5-hectare island rises only three metres above sea level and is made of seabird guano, sand and coral debris, all loosely held together by scant vegetation and a thin soil.

This cute little rodent was first sighted back to the 1800s by fishermen visiting the coral cay. Many years later, in 1978, researchers estimated several hundred rodents were thriving on the small remote island, yet over the next two decades their numbers declined rapidly. By 1998 they were down to double figures and, while researchers continued seeing individuals until 2009, it was obvious the melomys was critically endangered. In more recent years, extreme weather events and growing king tides in the Torres Strait had scientists worried. They noticed repeated episodes of severe seawater inundation and it was likely that the cay was almost completely submerged several times - it was probably one of these events that took the last of the animals. They also caused extensive erosion and there are estimates that the island has lost 90% of its vegetation cover in just a few decades.

Some scientists believe the Bramble Cay melomys could still exist on Papua New Guinea, although there is no historical evidence that this species has ever lived anywhere else. A cruel victim of geographic isolation and climate change, the melomys' story serves as a reminder of how important humanity's role will be in the the future preservation of our wildlife.

Many - perhaps hundreds of thousands - of species are, like the melomys, being slowly forced into a geographical last stand. Only successful efforts to restore habitats and reverse environmental damage will stand between them and certain extinction.

A SPECIES' LAST STAND
Hidden in the flotsam and jetsam of Bramble Cay, sheltered by clumps of grass and other small salt-resistant plants, the Bramble Cay melomys survived and thrived for what is likely to have been millennia. Isolated, but well adapted to the tiny island, it is known nowhere else in the world. It's now officially regarded as the first mammal extinction caused by climate change.

Woodslane Press Pty Ltd
10 Apollo Street
Warriewood, NSW 2102
Email: info@woodslanepress.com.au
Tel: 02 8445 2300
Website: www.woodslanepress.com.au

Published in Australia in 2023 by Woodslane Press

Text copyright © 2023 Myke Mollard
Illustrations copyright © 2023 Myke Mollard
The moral rights of the author and illustrator have been asserted.

This work is copyright. All rights reserved. Apart from any fair dealing for the purposes of study, research or review, as permitted under Australian copyright law, no part of this publication may be reproduced, distributed, or transmitted in any other form or by any means, including photocopying, recording, or other electronic or mechanical methods, without the prior written permission of the publisher. For permission requests, write to the publisher, addressed "Attention: Permissions Coordinator", at the address above.

Printed in China by Hang Tai
Designed by Myke Mollard
Special thanks to Dennis Jones and Ruthy Ramoso for diving into my world and giving me all your enthusiasm and support to help make this dream happen.

The information in this publication is based upon the current state of commercial and industry practice and the general circumstances as at the date of publication. Every effort has been made to obtain permissions relating to information reproduced in this publication. The publisher makes no representations as to the accuracy, reliability or completeness of the information contained in this publication. To the extent permitted by law, the publisher excludes all conditions, warranties and other obligations in relation to the supply of this publication and otherwise limits its liability to the recommended retail price. In no circumstances will the publisher be liable to any third party for any consequential loss or damage suffered by any person resulting in any way from the use or reliance on this publication or any part of it. Any opinions and advice contained in the publication are offered solely in pursuance of the author's and publisher's intention to provide information, and have not been specifically sought.

A catalogue record for this book is available from the National Library of Australia

MIX
Paper | Supporting responsible forestry
FSC® C023121

CHECK OUT MYKE'S GROWING LIBRARY OF FANTASTIC WILDLIFE TITLES

MARINE INSIGHT MONTHLY   GBR TIMES $1.75 EARTHDAY EDITION

GREAT BARRIER REEF TIMES

THE LATEST AND GREATEST NEWS AND CURRENT SCIENCE ON ALL THINGS GREAT AND SMALL WITH THE **GBR**.

THREATS TO THE REEF

Human impacts are playing a large role in upsetting nature's balance on the reef: climate change being just one of the threats to the creatures and environment. There are many things that affect the natural world, but many ways in which we can individually play a positive role in addressing them.

Here is a list of threats that are directly contributing to reef's degradation and loss. Simply thinking about the energy you use, food you eat and things you buy - and changing your habits to reduce your environmental impact - can make a small but crucial difference.

Global warming is indeed the biggest threat: warmer waters (and **ocean acidification**) result in **coral bleaching** events. Severe weather events such as **cyclones, flooding and storms**, are becoming more common because of increased energy in the atmosphere: 10 cyclones of category three or more crossed the Great Barrier Reef between 2004 and 2018, causing significant damage.

Increasing pollution and marine debris keep washing up: **microplastics** in the water are increasing and entering the food chain. Run-off and waste from **farming, agriculture, mining** and **urban coastal development** also increase water pollution. Mining - and **deforestation** - also unlock **natural pollutants** from the land which find their way to the sea and affect the reef.

Increased **shipping movements** mean more **vessel strikes** and **sound pollution** for whales and dugongs. **Groundings** and shallow reef crossings are becoming more frequent and can wipe out centuries of coral growth.

Increased **tourism** and recreational boating put oil residue, fuel and sunscreen into the water and create heavy tourism traffic with people ignorantly **walking on coral. Boat anchors** also further damage the reef.

Our love affair with fishing puts stress on the reef. **Overfishing**, game fishing, nets and lines tangled on reefs, and **unintentional bycatch** are all contribute to the growing list of threats to the reef.

The cumulative effect on the reef leads to **coral diseases, algae blooms and red tides.** These in turn can create opportunities for **invasive species,** such as the problematic crown of thorns starfish.

Within many of the illustrations in this book are small clues to the damage we are doing to the Great Barrier Reef. These 23 small but significant images are depictions of pollution and other threats to wildlife. A prize of a full set of my published books will go to the first person who writes into my publishers, Woodslane Press, with a full list of what and where they are. Good luck and happy hunting!

"Australians generate more single-use plastic waste per person than any other country in the world."

TOURISTS BEWARE

The Great Barrier Reef is a huge draw card for Australian tourism and Queensland's weather is beautiful one day, perfect the next! However, there are some real dangers lurking on the reef that many visitors are unaware of until they dive in unprepared. From Airlie Beach to the Torres Strait, a few of the biggest threats are the saltwater crocodiles, tiny Irukandji jellyfish and the infamous box jellyfish. Within many of the rockpools and in the tidal shallows there are deadly coneshells, stonefish and the highly venomous blue ringed octopus to watch out for.

Snorkeling and diving on the many reefs you could also encounter bull sharks, sea snakes and even red scorpionfish, butterfly cod and stingrays - and don't upset a giant Queensland grouper! On land there are coastal taipans and tiger snakes, and in the rainforest Cassowaries and stinging plants. But don't let this put you off! The vast majority of visitors have a perfectly safe stay. It's a wild place to explore, so be prepared for an amazing adventure!

WWW.WOODSLANEPRESS.COM.AU

THE FISH CLUSTER

1. Garfish, 2. Striped marlin, 3. Red emperor (striped morph), 4. Barracuda, 5. Sailfish, 6. Luddrick or banded blackfish, 7. Ocean bream or yellowfin snapper, 8. Australian snapper, 9. Sunfish, 10. Common yellow-fin tuna, 11. Mulloway or jewfish, 12. Kingfish or yellow-tailed kingfish, 13. Giant trevally, 14. Juvenile golden trevally or yellow pilot fish, 15. Sweetlips or coral snapper, 16. Pilot fish or black and white banned pilot fish, 17. Barramundi, 18. Mangrove jack, 19. Spanish mackerel, 20. Chinaman fish, 21. Queensland grouper, 22. Potato cod, 23. Coral trout, 24. Barramundi cod, 25. Stripey sea perch, spanish flag or blue banded hussar, 26. Brown striped sea perch or brown hussar, 27. Five-lined sea perch, 28. Red emperor, 29. Reef triggerfish, 30. Lagoon triggerfish, 31. Fingerspot snapper, fingermark sea perch or moses snapper, 32. Anchor tuskfish, 33. Red gurnard, 34. Black spotted yellow pufferfish, 35. Common reef pufferfish, 36. Stars and stripes pufferfish, 37. Sunset wrasse, 38. Striped reef trumpetfish, 39. Yellow trumpetfish, 40. Longfinned black banded batfish, 41. Dolphinfish (mahi mahi), 42. Palette surgeonfish, blue tang or Pacific blue tang, 43. Rainbow parrotfish, 44. Maori wrasse, moorhead wrase, colossal humphead wrasse or Napoleon wrasse, 45. Clown triggerfish, 46. Harlequin tuskfish, 47. Black banded pennant fish or long-finned bannerfish, 48. Butterfly cod, lionfish or scorpionfish, 49. Titan triggerfish, 50. Steephead parrotfish, 51. Green or bullethead parrotfish, 52. Emperor angelfish, 53. Silver batfish or silver moony, 54. Unicornfish, 55. Barcheek unicornfish, naso tang, or orange-spine unicornfish, 56. Orange damselfish or garibaldi damselfish, 57. Yellow damselfish, 58. Blue-spotted rabbitfish, 59. Blue damselfish, 60. Eyespot damselfish, 61. Bluespot yellow butterflyfish, 62. Tail spot wrasse and Hoeven's wrasse, 63. Humbug damselfish or black and white banded humbug, 64. Red scorpionfish or red scorpion rock cod, 65. Copperband butterflyfish, 66. Pacific double-saddled butterflyfish, 67. Longnosed yellow butterflyfish, 68. Bi-colour angelfish, 69. Brown double banded anemone fish or bi-banded clownfish, 70. Clownfish, orange and white banded anemonefish or Nemo fish, 71. Bumphead parrotfish, 72. Stripey surgeonfish, 73. Moorish idol, 74. Blue parrotfish.

THE CORAL CLUSTER

1. Elkhorn coral, 2. Seafan or gorgonian coral fan, 3. Plate coral or vase coral, 4. Seawhips or seaplumes, 5. Lettuce leaf coral or cabbage leaf coral, 6. Honeycomb coral, 7. Stove-pipe sponge, yellow tube sponge or cigar-like branching vase sponge, 8. Purple hump coral, 9. Weeping coral, long polyp toadstool leather coral colony, 10. Caepitose pink acropora branching coral, 11. Tabular acropora coral, 12. Carnation coral or tree coral, 13. Smooth cauliflower coral, 14. Grooved brain coral, 15. Giant barrel sponges, 16. Encrusted colony of bubble coral, 17. Clubbed finger coral or finger coral, 18. Massive cauliflower coral with encrusting green carpet sponge, 19. Arborescent staghorn coral, 20. Finger sponge and row-pore rope sponge, 21. Red organpipe coral, 22. Looking like coral seaplumes or seawhips, this is actually the giant noble featherstar, not unlike the brittle stars (free living coral flowers of the reef), 23. Brown tube sponge or branching tube sponge, 24. *Corymbose acropora* coral, 25. Giant carpet coral anemone, 26. Digitate acropora coral, 27. Massive starlet coral, 28. Encrusted Juvenile leaf coral, 29. Tabular leaf coral, 30. Finger leather coral tree, 31. Mushroom coral (Another free living species), 32. Blue flower-pot coral, 33. Sun coral, 34. Encrusting sponge, encrusting octopus sponge or carpet sponge.

REEF FISH

1. Red emperor (striped morph), 2. Red emperor, 3. Juvenile yellow-tailed kingfish, 4. Remora, 5. Mangrove jack, 6. Harlequin tuskfish, 7. Ocean mullet, 8. Coral trout, 9. Barramundi, 10. Barracuda, 11. Fingerspot snapper, fingermark sea perch or moses snapper, 12. Stripey sea perch, spanish flag or blue banded hussar, 13. Brown striped sea perch or brown hussar, 14. Five-lined sea perch, 15. Australian snapper, 16. Sweetlips or coral snapper, 17. Potato cod, 18. Chinaman fish, 19. Barcheek unicornfish, naso tang, or orange-spine unicornfish, 20. Butterfly cod, lionfish or scorpionfish, 21. Red sea bass, 22. John dory or St Peter's fish, 23. Leatherjacket or filefish, 24. Reef triggerfish, 25. Barramundi cod, 26. Palette surgeonfish, blue tang or Pacific blue tang, 27. Stripey surgeonfish, 28. Yellow trumpetfish, 29. Striped reef trumpetfish, 30. Silver batfish or silver moony, 31. Black banded pennant fish or long-finned bannerfish, 32. Anchor tuskfish, 33. Queensland grouper, 34. Moorish idol, 35. Surge wrasse, 36. Rainbow parrotfish, 37. Sunset wrasse, 38. Bumphead parrotfish, 39. Blue parrotfish, 40. Maori wrasse, moorhead wrase, colossal humphead wrasse or Napoleon wrasse, 41. Green or bullethead parrotfish, 42. Longfinned black banded batfish, 43. Orange damselfish or garibaldi damselfish, 44. Red squirrelfish, 45. Blue damselfish, 46. Yellow damselfish, 47. Drummerfish, 48. Black drummerfish, 49. Eyespot damselfish, 50. Copperband butterflyfish, 51. Bluespot yellow butterflyfish, 52. Black banded royal butterflyfish, 53. Longnosed yellow butterflyfish, 54. Yellow butterflyfish, 55. Bi-colour angelfish, 56. Regal angelfish, 57. Pacific double-saddled butterflyfish, 58. Emperor angelfish, 59. Humbug damselfish or black and white banded humbug, 60. Clown triggerfish, 61. Lagoon triggerfish, 62. Stars and stripes pufferfish, 63. Banned morwong, 64. Titan triggerfish, 65. Unicornfish, 66. Clownfish, orange and white banded anemonefish or Nemo fish, 67. Blue-spotted rabbitfish, 68. Red scorpionfish or red scorpion rock cod, 69. Stonefish, 70. Red gurnard, 71. Spotted flounder or flatfish, 72. Eastern blue devil, 73. Brown double banded anemone fish or bi-banded clownfish, nannygai or big-mouthed eastern nannygai, 75. Common reef pufferfish, 76. Black spotted yellow pufferfish, 77. Sea stars, 78. Seahorses, 79. Steephead parrotfish, 80. Tail spot wrasse and Hoeven's wrasse, 81. Sand flathead, 82. Reef butterflyfish, 83. Luddrick or banded blackfish.

OCEAN BIRDS

1. Brown Noddy, 2. Black Noddy, 3. Pelican, 4. Common Gull or Silver Gull, 5. Osprey, 6. Wandering Albatross, 7. Shearwaters: a range of different species are represented within this small window above the waves.

OCEAN AND PELAGIC FISH

8. Garfish, 9. Juvenile yellow-tailed kingfish, 10. Ocean mullet, 11. Dolphinfish (mahi mahi), 12. Common yellow-fin tuna, 13. Striped marlin, 14. Black marlin, 15. Kingfish or yellow-tailed kingfish, 16. Sailfish, 17. Giant trevally, 18. Juvenile golden trevally or yellow pilot fish, 19. Spanish mackerel, 20. Common Pacific mackerel, 21. Remora, 22. Australasian snapper, 23. Ocean bream or yellowfin bream, 24. Silver mulloway or jewfish, 25. Pilot fish or black and white banned pilot fish, 26. Sunfish, 27. Barracuda, 28. Sweetlips or coral snapper, 29. Luddrick or banded blackfish, 30. Australian salmon, 31. Wahoo.

SHARKS

1. Tiger shark, 2. Bronze whaler shark, 3. Hammerhead (common hammerhead and greater hammerhead), 4. Whaleshark, 5. Bull shark (related to the bronze whaler sharks, but a more robust species that hunts in mangroves, brackish waterways, up rivers, estuaries, marinas, Goldcoast canels and fringe reefs surrounding any of the 900 islands dotted across the Great Barrier Reef. 6. Blacktip reef shark, 7. Oceanic white tip shark, 8. Grey reef shark, 9. Small oceanic blue shark, 10. Thresher shark, 11. Leopard shark, 12. Grey nurse shark, 13. juvenile golden trevally or yellow pilot fish (can often seek the protection of oceanic sharks or large predatory fish, becoming a friendly entourage). 14. Black banded pilot fish, look like they are piloting their monsterous companions through the oceans. 15. Remora fish or sucker fish are pelagic oceanic fish that are often seen hitchhiking on whalesharks, tiger sharks, sea turtles or manta rays. 16. juvenile yellow-tailed kingfish often hang around or even school near large sharks. Like adult cobia, trevally and kingfish they seek protection when travelling in smaller groups by shadowing larger marine buddies.